Life Sciences Contribution
Royal Ontario Museum 97

A Revision of the Longipinnate Ichthyosaurs of the Lower Jurassic of England, with Descriptions of Two New Species (Reptilia: Ichthyosauria)

C. McGowan

ROM

LIFE SCIENCES CONTRIBUTIONS
ROYAL ONTARIO MUSEUM
NUMBER 97

C. McGOWAN

A Revision of the Longipinnate Ichthyosaurs of the Lower Jurassic of England, with Descriptions of Two New Species (Reptilia: Ichthyosauria)

Publication date: 29 April 1974
ISBN 0-88854-151-1
Suggested citation: Life Sci. Contr., R. Ont. Mus.

ROYAL ONTARIO MUSEUM
PUBLICATIONS IN LIFE SCIENCES

The Royal Ontario Museum publishes three series in the Life Sciences:

LIFE SCIENCES CONTRIBUTIONS, a numbered series of original scientific publications, including monographic works.

LIFE SCIENCES OCCASIONAL PAPERS, a numbered series of original scientific publications, primarily short and usually of taxonomic significance.

LIFE SCIENCES MISCELLANEOUS PUBLICATIONS, an unnumbered series of publications of varied subject matter and format.

All manuscripts considered for publication are subject to the scrutiny and editorial policies of the Life Sciences Editorial Board, and to review by persons outside the Museum staff who are authorities in the particular field involved.

CHRISTOPHER MCGOWAN is Assistant Curator, Department of Vertebrate Palaeontology, Royal Ontario Museum, and Assistant Professor, Department of Zoology, University of Toronto.

PRICE: $1.50

100 Queen's Park, Toronto, Canada

PRINTED AT THE BRYANT PRESS LIMITED

A Revision of the Longipinnate Ichthyosaurs of the Lower Jurassic of England, with Descriptions of Two New Species (Reptilia: Ichthyosauria)

Abstract

A revision of the longipinnate ichthyosaurs of the English Lias (Lower Jurassic) establishes five valid species. One of these is referred to the Upper Liassic genus *Stenopterygius*, the others to the genus *Temnodontosaurus* Lydekker, 1889, which is predominantly Lower Liassic in age. Two new Lower Liassic species are described. A detailed account of the variates measured is given.

Zusammenfaasung

Eine Revision der longipinnaten Ichthyosaurier des englischen Lias (Untere Jura) stellt fuenf gueltige Arten auf. Eine von diesen bezieht sich auf die aus dem oberen Lias stammende Gattung *Stenopterygius*, die anderen auf die Gattung *Temnodontosaurus* Lydekker, 1889, welche ueberwiegend aus dem unteren Lias stammt. Ein ausfuehrlicher Bericht ueber die gemmessenen Verschiedenheiten wird gegeben.

Introduction

The Lower Jurassic deposits of England are well endowed with the remains of ichthyosaurs, which have been collected intensively since the beginning of the last century. Most material came from the Lower Lias [Hettangian and Sinemurian], the principal localities being Lyme Regis, Dorsetshire; Street, in the adjoining county of Somersetshire; and Barrow-on-Soar, Leicestershire. Ichthyosaurs are also known from the Upper Lias [Lower Toarcian], but this horizon has so far yielded only longipinnates. The principal Upper Liassic locality is the Whitby district of Yorkshire, equivalent in age to the famous Holzmaden locality of Germany. Longipinnates are also found in the Lower Lias but are less numerous than latipinnates. The distinction between latipinnates and longipinnates is discussed in detail elsewhere (McGowan, 1972a), the major differences occurring in the forefins; longipinnates have three distal carpals hence three primary digits, whereas latipinnates have four distal carpals thus four primary digits.

A recent review of the Lower Liassic latipinnates (McGowan, in press) showed the systematics to be confused, two thirds of the species being invalid. Similar confusion occurs in the longipinnates and the problem is complicated by the question of the generic name. The Liassic latipinnates belong to the genus *Ichthyosaurus* De La Beche and Conybeare, 1821, but the appropriate name for the Liassic longipinnates is problematic. Charig (in preparation) has made a detailed study of the nomenclatural status of the English Liassic ichthyosaurs. He notes that as *Temnodontosaurus* Lydekker, 1889 (in Nicholson and Lydekker, 1889) has *I. platyodon* Conybeare, 1822, as its type-species, *Temnodontosaurus* is available as the generic name for that taxon. My study shows *I. platyodon* to be one of a group of four species referable to a single genus, and *Temnodontosaurus* thus has priority as the generic name for these longipinnates.

As many as eight longipinnate species have been described from the English Lias (see Lydekker, 1889a; Von Huene, 1922) and these will be referred, for the present, to their original genus, *Ichthyosaurus*: *I. acutirostris* Owen, 1840; *I. crassimanus* Blake, 1876; *I. latifrons* Koenig, 1825; *I. lonchiodon* Owen, 1840; *I. longifrons* Owen, 1881; *I. longirostris* Mantell, 1851; *I platyodon* Conybeare, 1822; and *I. zetlandicus* Seeley, 1880. The purpose of this paper is to clarify the systematics of these English longipinnates.

Materials and Methods

Materials

Measurements were taken directly from 20 specimens and from photographs of an additional four specimens. Of these, 13 were from the Lower Lias [Hettangian and Sinemurian] of Lyme Regis, Street, and Barrow-on-Soar, and 11 from the Upper Lias [Lower Toarcian] of Whitby. The abbreviations used for collections examined are: BMNH, British Museum (Natural History), London; LCM, Leicester County Museum, England; OUM, Oxford University Museum; ROM, Royal Ontario Museum, Toronto; SCC, Street Collection, in the care of Clarks Shoe Company, Street, Somerset; SMC, Sedgwick Museum, Cambridge University; UW, University of Wyoming; WMY, Whitby Museum, Yorkshire; and YM, York Museum, Yorkshire.

Methods

All measurements were made with vernier calipers accurate to .01 cm, except those of body length which were made with a tape accurate to 0.1 cm. Because of incomplete preservation or exposure of bony elements (the term *indeterminate* is used in this context in the sections which follow) few post-cranial characters were used, viz: length of body, vertebral count to pelvic girdle, vertebral count to tail bend, length of forefin, width of forefin, total digital count, number of primary digits, number of accessory digits, number of elements in longest digit, and occurrence of notching (Fig. 1).

The cranial measurements recorded are shown in Fig. 2. Because of size variation it was necessary to convert all continuous variates into ratios, and those derived from cranial measurements are: diameter of orbit (3) to length of jaw (7), length of snout (6) to length of jaw (7), length of premaxillary segment (1) to length of jaw (7), length of

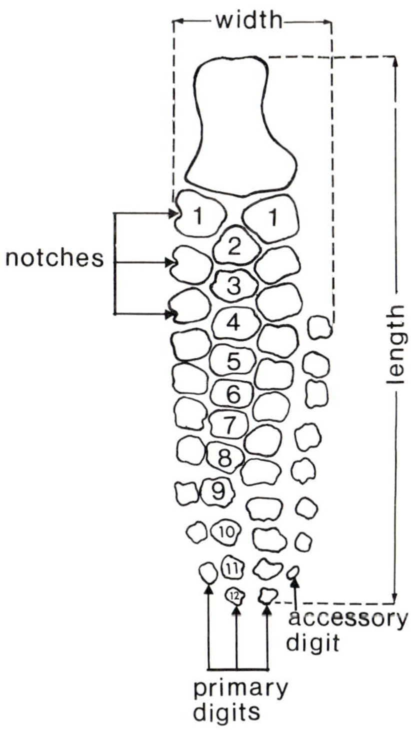

Fig. 1 Diagram of a longipinnate forefin (left, dorsal view) showing the distinction between primary and accessory digits. The maximum number of elements in the longest digit is 12.

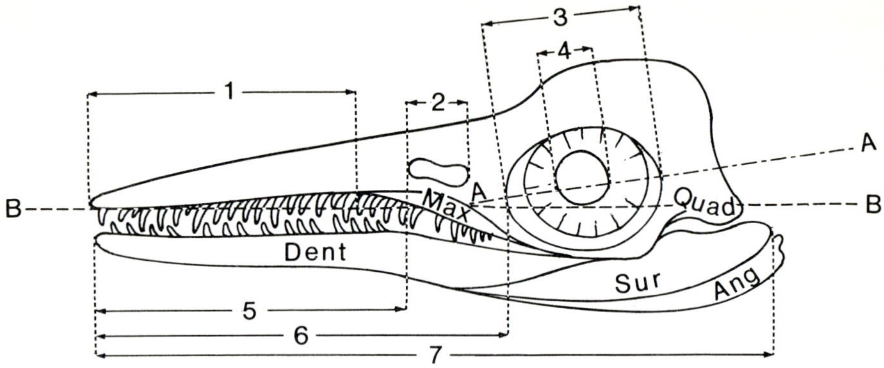

Fig. 2 Diagram of an ichthyosaur skull showing the cranial measurements: 1, length of premaxillary segment; 2, length of external naris; 3, diameter of orbit; 4, internal diameter of sclerotic ring; 5, length of prenarial segment; 6, length of snout; 7, length of jaw.
A-A, longitudinal axis of orbit; B-B, longitudinal axis of skull; Ang, angular; Dent, dentary; Quad, quadrate; Sur, surangular; Max, maxilla.

prenarial segment (5) to length of jaw (7), length of external naris (2) to diameter of orbit (3), internal diameter of sclerotic ring (4) to diameter of orbit (3). Due to small sample sizes and missing data a statistical analysis was not possible. It is therefore not known whether the ratios are affected by allometric growth, nor whether they are normally distributed. However, I have found intercranial growth (relative to jaw length) to be almost completely isometric in *Stenopterygius quadriscissus*, *Ichthyosaurus communis* and *I. breviceps* (McGowan, 1973a) and this might also hold for the present species. Furthermore an analysis of data for *I. communis* and *I. breviceps* (McGowan, in press) and *S. quadriscissus* (McGowan, in preparation) has shown all ratios to be normally distributed.

Details of Variates Recorded

AXIAL SKELETON

Length of body: the length, measured along the vertebral column, from first to last vertebra.

Vertebral count to the pelvic girdle: the number of vertebrae counted from the atlas to the position of the pelvic girdle. The atlas is quite distinct because it is fused with the axis (Fig. 3A). The position of the pelvic girdle is taken as that centrum which lies closest to the vertical level of the ilium. The pelvic girdle is frequently displaced or lost and in this case an estimate is made based upon the position of the proximal head of the femur.

Vertebral count to the tail bend: the number of vertebrae counted from the atlas to the apex of the tail bend. Where the tail bend is distinct there is little difficulty in making the count which is continued up to, and includes, that centrum at the apex of the bend. The apical centra are

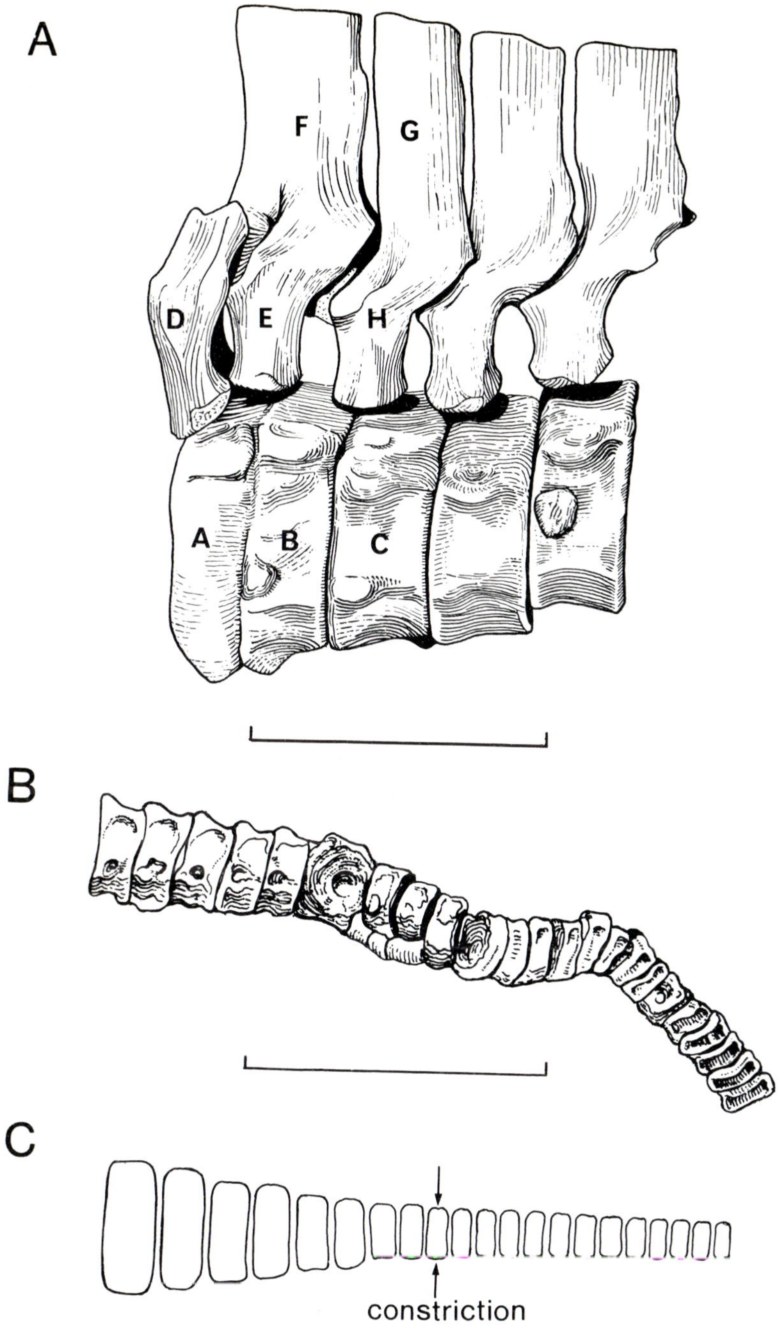

Fig. 3 Features of the ichthyosaurian vertebral column, scale = 10 cm.

A. Anterior portion of vertebral column of *Ichthyosaurus communis*, Lower Lias, Lyme Regis; A, atlas centrum; B, axis centrum; C, centrum of third cervical vertebra; D, neural arch of atlas; E, neural arch of axis; F, fused neural spine of atlas and axis; G, neural spine of third vertebra; H, neural arch of third vertebra.

B. Tail bend region of *Ichthyosaurus communis* (BMNH 2013), Lower Lias, Lyme Regis. Note the wedge-shaped centra at the apex of the bend.

C. Diagrammatic representation of the tail bend region in a dorso-ventrally preserved specimen showing how the rate of decrease in diameter of the centra is reduced distal to the apex of the bend.

wedge-shaped, widest dorsally, and these clearly mark the apex of the tail bend (Fig. 3B). Where the bend is indistinct, as in dorso-ventrally compressed specimens, its position can often be estimated by considering the rate of taper in the diameter of the centra. The gradual decrease in diameter towards the tip of the tail is less marked in those centra lying distal to the tail bend than in those proximal to the tail bend. Consequently there is a slight constriction at this level (Fig. 3C). When a constriction is not apparent its position can often be revealed by measuring the diameters of the centra, a marked difference in the diameters of adjacent centra indicating the position of the tail bend.

SKULL (see Fig. 2)

Length of skull: the distance between the tip of the snout and the posterior edge of the articular surface of the quadrate. The quadrate is selected as a point of reference because it has no contact with post-cranial structures and is therefore usually well defined. Occasionally the quadrate is displaced, in which case allowances must be made.

Length of snout: the distance between the tip of the snout and the anterior boundary of the orbit. Von Huene (1922) measured to the anterior boundary of the external narial opening, but the boundaries of the naris are often indeterminate.

Length of prenarial segment: the distance between the tip of the snout and the anterior boundary of the external naris. Occasionally the bones surrounding the narial aperture are pushed inwards, forming a triangular depression, and this may be misinterpreted as the external naris. As a general guide the anterior boundary of the external naris is rounded, whereas depressions tend to be angular.

Length of premaxillary segment: the distance between the tip of the snout and the anterior tip of the maxilla. The free margin of the maxilla forms a continuous straight edge with that of the premaxilla, and a break in continuity in this region is indicative of damage or displacement of the maxilla. Where such damage has occurred an approximation can be made by projecting the maxillary margin forward until it intersects with the premaxilla.

Longitudinal diameter of orbit: the maximum internal diameter of the orbit. The orbit is oval rather than round, and its longitudinal axis may not necessarily correspond with the longitudinal axis of the skull (an axis passing through the tip of the snout and the articular surface of the quadrate). The longitudinal diameter of the orbit is measured along the orbital axis, at the widest point. The posterior boundary of the orbit is invariably well defined and is the anterior surface of the postorbital. The anterior boundary is the internal (posterior-facing) surface of the lachrymal. Sometimes the lachrymal bears a posterior projection of bone which has the appearance of a crescentic swelling in the antero-dorsal corner of the orbit and which interferes with the measurement. However, by projecting the line of the orbit (i.e., the curvature of the jugal) forward, it is possible to estimate the anterior boundary of the orbit.

Internal diameter of sclerotic ring: the maximum internal diameter of the sclerotic ring. The aperture of the sclerotic ring is usually oval, and the maximum internal diameter lies approximately parallel to the longitudinal axis of the orbit. If distortion has occurred and the maximum diameter lies perpendicular to the longitudinal axis of the orbit, or if the aperture is asymmetric, a correction must be made. In both instances the internal diameter is taken as the mean of the maximum diameter, and the diameter measured at right angles to it.

Length of external naris: the distance between anterior and posterior boundaries of the external naris measured between verticals.

LOWER JAW

Length of jaw: the distance between the anterior tip of the dentary and the posterior edge of the surangular, measured between verticals perpendicular to the longitudinal axis of the skull.

DENTITION

Form of teeth: the teeth lie with their bases in a dentigerous groove and for this reason are seldom seen in their entirety. Note is taken of the tooth shape.

Number of maxillary teeth: the number of teeth in the upper jaw counted from the most posterior tooth, to that at the anterior tip of the maxilla. The most anterior maxillary tooth may lie partly in contact with the premaxilla, but provided it maintains contact with the maxilla it is counted as a maxillary tooth.

FINS (see Fig. 1)

Length of fore- and hindfin: the distance between the distal end of the propodial and the most distal phalanx, measured between horizontals perpendicular to the longitudinal axis of the fin. The propodial is not included in the length of the fin because of the difficulty in accurately determining its length. The terminal phalanges are usually very small and often well spaced so that they do not make contact with one another. Sometimes they are obscured by matrix, or lost altogether, but if the most distal phalanges are relatively small it may be assumed that the fin is complete.

Width of fore- and hindfin: the maximum width measured between verticals parallel to the longitudinal axis of the fin.

Number of primary digits: the maximum number of primary digits.

Total digital count: the total number of primary and accessory digits. Accessory digits frequently contain only a few elements, but if they exceed three in number they are counted.

Occurrence of notching in fin elements: notches are often obscured by matrix, which it is sometimes necessary to remove.

Number of elements in longest digit: numerically equal to the horizontal level of the most distal fin element. The count commences at the level of the epipodials (level one).

Estimates of Growth Constants α and b

Estimates of the allometric growth constants α and b for the growth of the head relative to the body were evaluated using the logarithmic transformation of the allometric growth equation $y = bx^{\alpha}$ (where $y =$ organ size and $x =$ body size). Bartlett's Model II regression technique was used in the computation, but the small sample size did not permit confidence limits to be set for the estimates of α and b (for details of the method see McGowan, 1973a).

Systematic Descriptions

I find that only three of the eight previously described English longipinnate species from the Lower Jurassic are valid; *I. platyodon* Conybeare, *I. acutirostris* Owen and *I. longirostris* Mantell. To these three are added two new species (described below), both from the Lower Lias of Lyme Regis, Dorset. *I. crassimanus* Blake, *I. longifrons* Owen and *I. zetlandicus* Seeley are all junior synonyms for *I. acutirostris*; *I. lonchiodon* Owen is a junior synonym for *I. platyodon,* and *I. latifrons* Koenig is rejected as a *nomen dubium.*

I. acutirostris (Upper Lias, Yorkshire) has closer affinity with the German species *Stenopterygius quadriscissus* (Quenstedt) (Upper Lias) than with the other English longipinnates, and is accordingly referred to the genus *Stenopterygius.* The remaining four English species are referred to the genus *Temnodontosaurus* Lydekker (1889b, pp. xi, 1126).

Class Reptilia
Subclass Euryapsida(?)
Order Ichthyosauria
Suborder Longipinnati Von Huene, 1948

Family *Temnodontosauridae* fam. nov.

Diagnosis

Forefin with no more than three primary digits, one of which supported by intermedium; total digital count probably does not exceed four; orbit relatively small, ratio, diameter of orbit to length of jaw $\leqslant 0.21$; aperture of sclerotic ring to diameter of orbit probably always < 0.35; maxilla relatively long, ratio, length of premaxillary segment to length of jaw $\leqslant 0.40$; external naris frequently relatively long, ratio, length of external naris to diameter of orbit often > 0.45; moderate to very large, adult jaw length $\geqslant 60$ cm and usually > 100 cm. Liassic in age and predominantly from the Lower division.

Type genus *Temnodontosaurus* Lydekker 1889b

Diagnosis

As for the Family Temnodontosauridae

Type species

Temnodontosaurus platyodon (Conybeare)

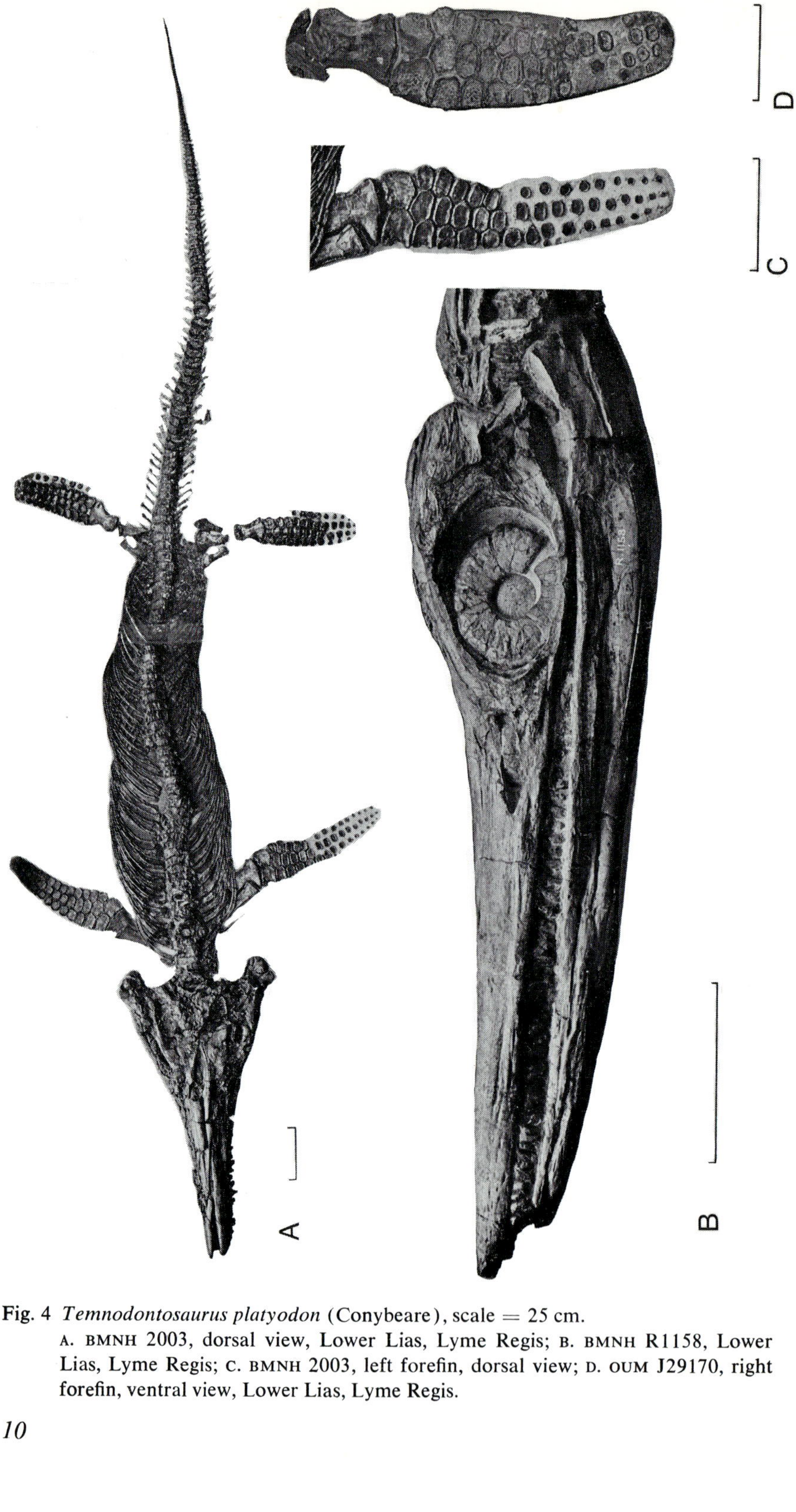

Fig. 4 *Temnodontosaurus platyodon* (Conybeare), scale = 25 cm.
A. BMNH 2003, dorsal view, Lower Lias, Lyme Regis; B. BMNH R1158, Lower Lias, Lyme Regis; C. BMNH 2003, left forefin, dorsal view; D. OUM J29170, right forefin, ventral view, Lower Lias, Lyme Regis.

Temnodontosaurus platyodon (Conybeare)
Fig. 4

Ichthyosaurus platyodon Conybeare, 1822, p. 108, pl. 15, fig. 7; pl. 16, figs. 1-7.
Ichthyosaurus platyodon Conybeare, De La Beche, 1826, p. 27.
Ichthyosaurus chiroligostinus Hawkins, 1834, p. 14, pl. 3.
Ichthyosaurus chiroligostinus Hawkins, Hawkins, 1840, p. 10, pl. 2, pl. 3.
Ichthyosaurus platyodon Conybeare, Owen, 1840, p. 112.
Ichthyosaurus lonchiodon Owen, 1840, p. 116.
Ichthyosaurus platyodon Conybeare, Mantell, 1851, p. 380.
Ichthyosaurus lonchiodon Owen, Mantell, 1851, p. 384.
Ichthyosaurus platyodon Conybeare, Owen, 1881, p. 115, pl. 24, fig. 4; pl. 31, figs. 1-3.
Ichthyosaurus lonchiodon Owen, Owen, 1881, p. 117, pl. 24, fig. 6; pl. 31, figs. 4-7.
Ichthyosaurus platyodon Conybeare, Lydekker, 1889a, p. 94, fig. 34.
Ichthyosaurus lonchiodon Owen, Lydekker, 1889a, p. 92.
Temnodontosaurus platyodon (Conybeare), Lydekker, 1889b, p. xi.
Leptopterygius platyodon (Conybeare), Von Huene, 1922, p. 18, pl. 2, fig. 2; pl. 3, fig. 3.
Leptopterygius lonchiodon (Owen), Von Huene, 1922, p. 16, pl. 3, fig. 1.
Leptopterygius (Ichthyosaurus) platyodon (Conybeare), Delair, 1969, p. 10.
Leptopterygius platyodon (Conybeare), Delair, 1969, p. 11.
Proteosaurus platyodon (Conybeare), McGowan, 1972a, p. 5, fig. 2b.
Proteosaurus platyodon (Conybeare), McGowan, 1972b, p. 5, figs. 9, 11.

Emended Diagnosis

Very large ichthyosaurs, length of jaw $>$ 100 cm, and $>$ 150 cm in mature individuals; snout relatively long and straight, ratio, length of snout to length of jaw $\geq$ 0.62; skull and lower jaw with rectilinear profile; orbit relatively small, ratio, diameter of orbit to length of jaw not exceeding 0.18 and probably always $<$ 0.18; sclerotic aperture relatively small, ratio, internal diameter of sclerotic ring to diameter of orbit $<$ 0.35; maxilla relatively long, ratio, length of premaxillary segment to length of jaw $<$ 0.36; external naris relatively long, ratio, length of external naris to diameter of orbit usually $>$ 0.60; ratio, length of prenarial segment to length of jaw $>$ 0.42; forefin long and narrow with three primary digits and probably one post-axial accessory digit; maximum number of elements in longest digit does not exceed 18; radius and radiale notched, possibly other elements too. Lower Liassic in age [Upper Hettangian to Lower Sinemurian] and largely, if not wholly, confined to Lyme Regis, Dorsetshire.

Original type material

A single tooth was figured in the original description of the species *I. platyodon* (Conybeare, 1822, pl. 15, fig. 7), which was in the care of the Geological Society of London (Woodward and Sherborn, 1890, p. 240). A search for this tooth was unsuccessful (Delair, 1960).

Proposed Neotype

BMNH 2003, a complete skeleton in the British Museum (Natural History).

Evidence that proposed neotype is consistent with what is known of original type material

Conybeare's original description of *I. platyodon* made reference to the large size, "the most gigantic yet discovered" (1822, p. 108). The proposed neotype, BMNH 2003, with a total length of 6.38 metres, is one of the largest Lower Liassic specimens, and is therefore consistent with Conybeare's original description.

Evidence that proposed neotype came as nearly as practicable from original type-locality

Conybeare did not specify the type-locality but did note that the species occurred in the Lias (1822, p. 108). At that time most ichthyosaurian material was being collected from Lyme Regis, and to a lesser extent from Somerset. As both of these localities are Lower Liassic, it may be inferred that the material was Lower Liassic in age.

Other material

BMNH R2918, BMNH 14564 (complete skeletons); OUM J29170 (incomplete skeleton); BMNH R1158, BMNH R1155, BMNH R215 (complete or near complete skulls) and ROM 7972 (partial skull).

Locality and Horizon

All material thus far referred to this species is from the Lower Lias of Lyme Regis, Dorset. De La Beche (1822) noted that remains of ichthyosaurs were "by no means rare" and were principally discovered at Black Ven, most commonly in the slaty or marly part of the Lias. The lowest horizon exposed at the foot of Black Ven is that of *Arnioceras semicostatum* (Fig. 5). In a later paper De La Beche (1826) recorded that the Liassic limestones of Lyme Regis contain numerous organic remains, including ichthyosaurs, which are found principally in the marls alternating with limestones containing the ammonite, *Arietites bucklandi* (i.e., the top of the Blue Lias). Most of the material from Lyme Regis was probably collected from the Blue Lias, from the zones of *Arietites bucklandi* and *Arnioceras semicostatum* (see Arkell, 1933). While it is possible that ichthyosaurs have been collected from higher zones, the geological range of specimens from Lyme Regis will be taken to be from the zone of *Schlotheimia angulata* to that of *Arnioceras semicostatum* [Upper Hettangian–Lower Sinemurian].

Emended Description

Temnodontosaurus platyodon is one of the largest ichthyosaurs, second only to *Cymbospondylus petrinus* (Middle Traissic, Nevada) in total body

STAGES		ZONES	LITHOLOGICAL FORMATIONS	LOCALITIES
SINEMURIAN	UPPER	*Echioceras raricostatum*	Black Marl of Black Ven	Lyme Regis Dorset
		Oxynoticeras oxynotum		
		Asteroceras obtusum		
	LOWER	*Caenisites turneri*		
		Arnioceras semicostatum	Shales-with-"Beef"	
		Arietites bucklandi	Blue Lias	
HETTANGIAN		*Schlotheimia angulata*		
		Alsatites liasicus		
		Psiloceras planorobis		Street Somerset

Fig. 5 The ammonite zones of the English Lower Lias (based upon Dean *et al.*, 1961).

length. *C. petrinus* has been estimated to reach 10 metres (McGowan, 1972b), whereas *T. platyodon* probably did not exceed 9 metres (data from BMNH R1155 and BMNH 2003).

SKULL

Snout relatively long and straight, tapering gently to the tip. Ratio, length of snout to length of jaw 0.64 in neotype, observed range 0.62–0.64 (n = 3). Maxilla long and robust, makes substantial contribution to margin of upper jaw; ratio, length of premaxillary segment to length of jaw 0.34 in neotype, observed range 0.30–0.34 (n = 2). As noted elsewhere (McGowan, 1972a), possession of a relatively long maxilla is a primitive character.

Orbit relatively small; ratio, diameter of orbit to length of jaw 0.16 in the neotype, observed range 0.16–0.17 (n = 3). Sclerotic aperture also relatively small; ratio, internal diameter of sclerotic ring to diameter of orbit (indeterminate in neotype) 0.32 in ROM 7972, 0.31 in BMNH R1158 and 0.24 in BMNH R215. Possession of small orbital and sclerotic diameters are both primitive characters (McGowan, 1972a). External naris large; ratio, length of external naris to diameter of orbit, probably always exceeding 0.50, is 0.53 in ROM 7972, 0.63 in BMNH R1158 and BMNH R1155, and 0.72 in BMNH R215. Ratio, length of prenarial segment to length of jaw 0.50 in neotype, range 0.46–0.50 (n = 3). Naris bilobed in BMNH R1158 (Fig. 6B), but not bilobed in either ROM 7972 or BMNH 14564 and not therefore a constant character. External lateral surfaces of tips of premaxillae marked in neotype (Fig. 6A) by fairly deep longitudinal striations, which converge anterodorsally giving serrated medial margin.

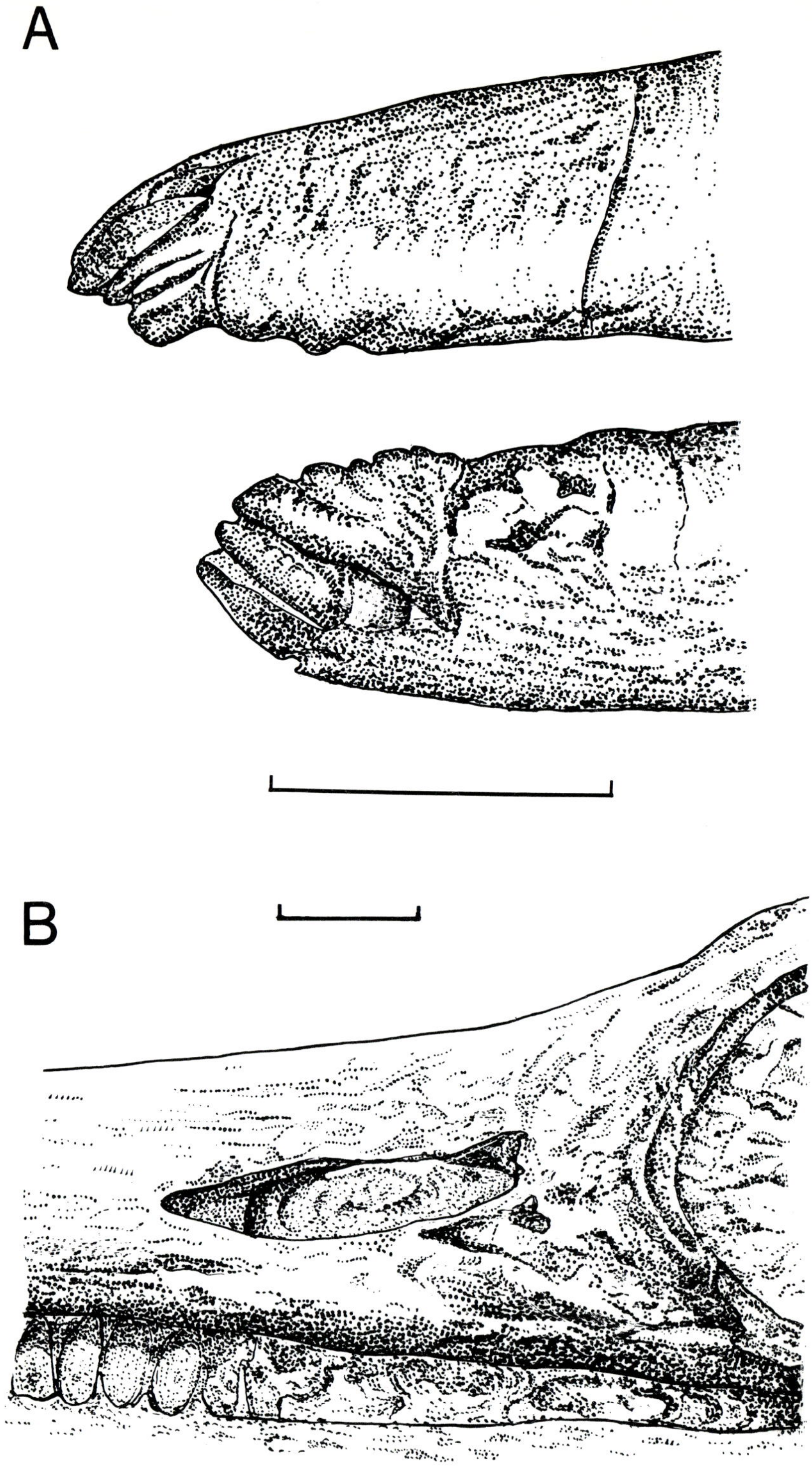

Fig. 6 *Temnodontosaurus platyodon* (Conybeare), scale = 5 cm.
A. BMNH 2003, tip of snout, dorsal view, showing striations.
B. BMNH R1158, external naris, lateral view, showing bilobed appearance.

Because extreme tip of snout frequently lost in other specimens, it is not known if this is a diagnostic character. Internasal foramen (McGowan, 1973b) present in neotype, BMNH R1155, BMNH R215, and ROM 7972, but absent in BMNH R1158 and indeterminate in remaining specimens; may be considered a constant character.

Probably also to be referred to *T. platyodon* are BMNH 2149 (and BMNH 2150), a basioccipital and basisphenoid with associated atlas and axis, and BMNH 481, a partial vertebral column. Both specimens are from the Lower Lias of Lyme Regis, and are large (dorso-ventral height of centra 18.5 cm, width of basisphenoid 24.7 cm). It is noteworthy that the basioccipital is not drawn out into a peg anteriorly, as in latipinnates (McGowan, 1973b) but has a flat anterior face. A similar basioccipital condition occurs in the Cretaceous longipinnate *Platypterygius americanus* (UW 2421), and in an undescribed longipinnate skull from the Upper Jurassic of Norfolk (SMC J68516). It is suggested that the absence of a basioccipital peg may be a longipinnate character.

FINS

Whereas the skull of *T. platyodon* is adequately known, knowledge of fin structure is incomplete because of scarcity of associated skeletons. Fore- and hindfins both preserved in neotype, but close inspection reveals right forefin is entirely plaster, and last two (distal) horizontal rows of left hindfin are plaster also. From level six onwards (radius and ulna are designated level one, radiale, intermedium and ulnare level two, and so on; see "Materials and Methods") elements of left forefin well spaced, giving suspiciously unnatural appearance. However, all elements are embedded in matrix (which has been painted) and there is even a small depression indicating where an element has been lost (distal to level 8); left forefin therefore genuine. Bones of two hindfins similarly embedded in matrix and likewise reliable.

In neotype hindfin approaches size of forefin; ratio, forefin length to hindfin length 1.4 (mean value 2.5 in *Stenopterygius quadriscissus*). Both fins long and narrow; aspect ratio (maximum length/maximum width) of left forefin 3.4 (mean value 2.3 in *S. quadriscissus*). Aspect ratio of left and right hindfins 2.6 and 2.2 respectively (mean value 2.1 in *S. quadriscissus*). Three primary digits in neotype forefin with evidence of a pre-axial accessory digit commencing at level eight; as this consists of only two elements it can be ignored. Maximum number of elements in the longest digit 16. Notching in first three pre-axial elements. Forefins incomplete in BMNH R2918 but have three primary digits, one post-axial accessory. Longest digit has 12 elements, but incomplete. In OUM J29170 right forefin more complete than left, has three primary and one accessory digits. Maximum number of elements in longest digit 14; radius and radiale notched.

In conclusion, the forefin has three primary digits and one post-axial accessory digit, the maximum number of elements in the longest digit probably does not exceed 18, and the radius and radiale are notched. The hindfin (in the neotype) has three primary and one post-axial accessory

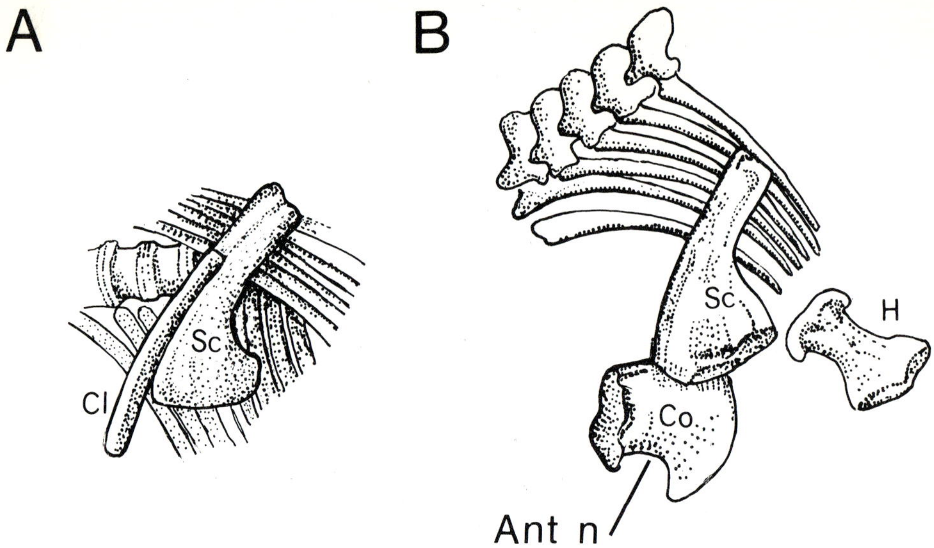

Fig. 7 A comparison of the pectoral girdle of BMNH 2003 and BMNH 14564 (after Owen, 1881).
Cl, clavicle; Co, coracoid; H, humerus; Sc, scapula; Ant n, anterior notch.
A. BMNH 2003, Lower Lias, Lyme Regis (pl. 31, fig. 3). [*I. platyodon*]
B. BMNH 14564, Lower Lias, Lyme Regis (pl. 31, fig. 4). [*I. lonchiodon*]

digits, and the maximum number of elements in the longest digit is 12. The forefin is only marginally longer than the hind, and has a high aspect ratio.

AXIAL SKELETON AND GIRDLES

Vertebral count from atlas to pelvic girdle 46 or 47 in neotype, estimated 46 in BMNH R2918, and 46 or 47 in BMNH 14564. Only in BMNH R2918 is vertebral column sufficiently well preserved for a vertebral count to the tail bend, estimated at 86 or 87 (mean counts to pelvic girdle and tail bend 47 and 83 respectively in *S. quadriscissus*, and 44 and 77 in the latipinnate *Ichthyosaurus communis*). Pre-flexural body segment relatively longer in *T. platyodon* than *S. quadriscissus* (see McGowan, 1972b, Fig. 17); *T. platyodon* was a longer-bodied animal. Individual vertebral centra with typical ichthyosaurian disc shape; in neotype (jaw length 148 cm) diameter of dorsal centra 11.5 cm, in OUM J29170 (jaw length 169 cm), dorsal centra 14.5 cm. In longest skull recorded, BMNH R1155 (jaw length approximately 221 cm), cervical centra 13.5 cm diameter compared with 18.5 cm in partial vertebral column BMNH 481. It is therefore concluded that the length of the jaw may have exceeded 221 cm. The limb girdles are invariably indeterminate in ichthyosaurs, and thus of reduced systematic interest, but it may be noted that they are relatively large and robust in *T. platyodon*. The coracoid apparently has only an anterior notch (Fig. 7B: Ant n).

DENTITION

Teeth are of little value in ichthyosaur classification (McGowan, 1969) and it is interesting to note that Conybeare's flattened "*platyodon tooth*" (Conybeare, 1822, pl. 15, fig. 7) is absent from all specimens thus far referred to the species *T. platyodon* (and for that matter from other ichthyosaurs examined). Teeth are clearly visible in the neotype (BMNH 2003), BMNH R1158, BMNH R215, and ROM 7972 and all are round in cross-section.

Remarks

Figures were not given by Owen in his description of *I. lonchiodon*, but reference was made to a specimen 15 feet (4.5 metres) long in the British Museum (BMNH 14564) purchased from T. Hawkins and figured in a subsequent paper (Owen, 1881, pl. 31, figs. 4-7). In his original description Owen compared *I. lonchiodon* with *I. platyodon* noting the following differences: head of *I. lonchiodon* relatively longer, lower jaw deeper and tapering more rapidly to tip, teeth more slender with circular cross-section, vertebral centra wider antero-posteriorly, scapula with deeper posterior concavity, forefin relatively smaller and fore-and hindfins do not approach similar size as they do in *I. platyodon*. Owen referred BMNH 2003 and BMNH R1158 to *I. platyodon* (1881, pl. 31, figs. 1 and 2 respectively), but if these specimens are compared with the holotype of *I. lonchiodon* (BMNH 14564) they are found to be indistinguishable on Owen's diagnostic characters:

	BMNH 2003	**BMNH R1158**	**BMNH 14564**
Length of skull / Length of body	0.28	—	0.28
Depth of jaw / Length of jaw	—	0.10	0.10
Shape of tooth section	Round	Round	—
Width of centrum / Height of centrum (30th centrum)	0.40	—	0.38

Furthermore, if the scapula of BMNH 2003 is compared with that of BMNH 14564 (Owen, 1881, pl. 31, figs. 3-4), it is seen that while they differ somewhat in shape, the posterior concavity is no deeper in BMNH 2003 (Fig. 7). The species *I. lonchiodon* Owen is thus concluded to be a junior synonym for *T. platyodon* (Conybeare).

Temnodontosaurus risor sp. nov.

Fig. 8

Etymology

Risor, Latin (masc.), a mocker

Ichthyosaurus communis Conybeare, Owen (*pro-parte*), 1840, p. 108.
Ichthyosaurus communis Conybeare, Lydekker (*pro-parte*), 1889a, p. 41, fig. 21.

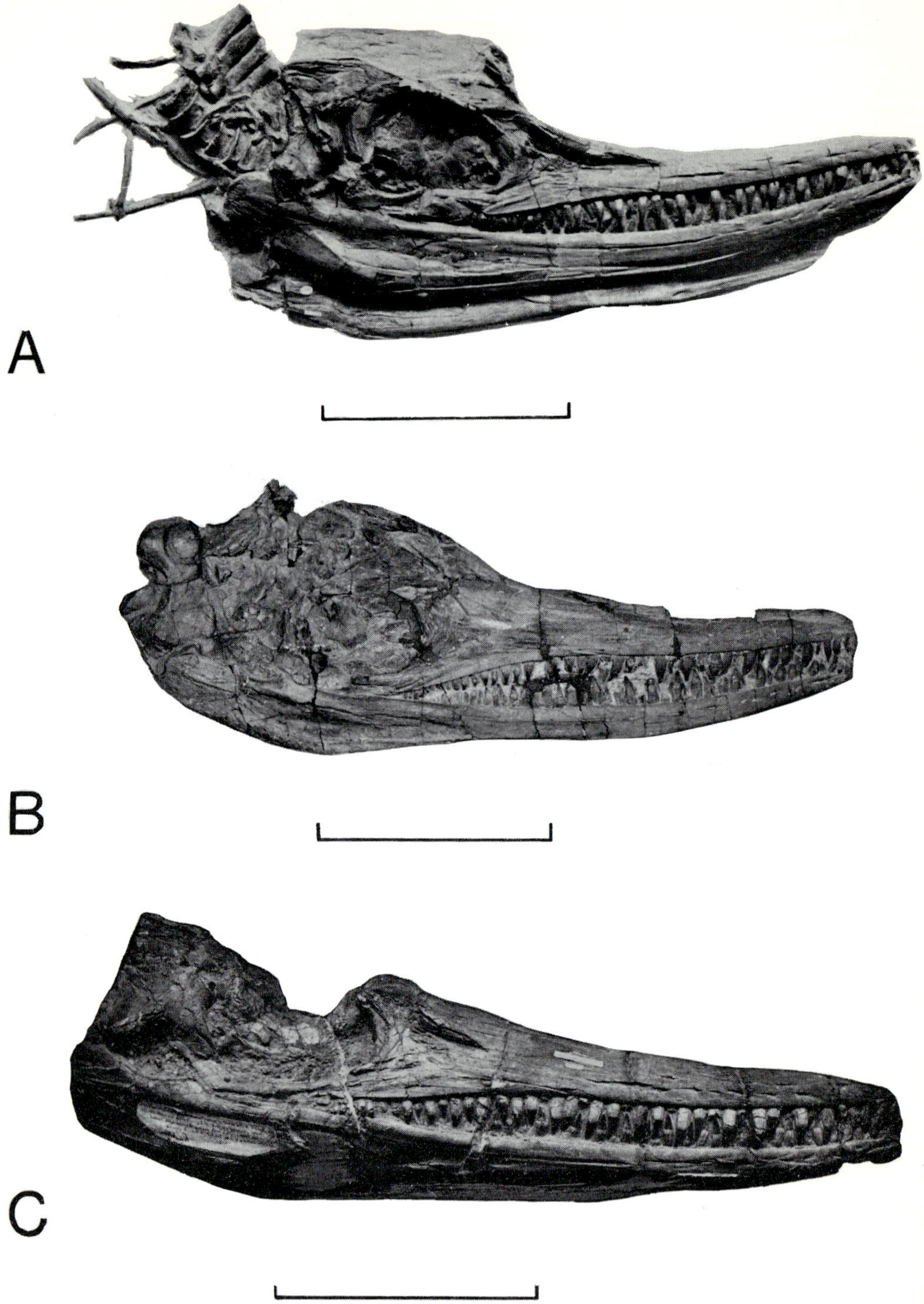

Fig. 8 *Temnodontosaurus risor* sp. nov., scale = 25 cm.
A. BMNH 43971, holotype, Lower Lias, Lyme Regis.
B. SMC J68446, Lower Lias, Lyme Regis.
C. BMNH R311, Lower Lias, Lyme Regis.

Diagnosis

Medium sized ichthyosaurs, length of jaw $\leqslant$ 100 cm; snout distinctly recurved, ratio, length of snout to length of jaw $>$ 0.57 but probably not exceeding 0.62; skull and lower jaw with recurved profile; orbit relatively large, ratio, diameter of orbit to length of jaw $\geqslant$ 0.18; sclerotic aperture relatively small, ratio, internal diameter of sclerotic ring to diameter of orbit $<$ 0.35; maxilla of moderate length, ratio, length of premaxillary segment to length of jaw $>$ 0.36; ratio, length of prenarial segment to length of jaw $>$ 0.42; maxillary tooth count $>$ 15. Lower Liassic in age [Upper Hettangian to Lower Sinemurian]; largely, if not wholly, confined to Lyme Regis, Dorsetshire.

Holotype

BMNH 43971, a complete skull in the British Museum (Natural History).

Other Material

SMC J68446 (complete skull); BMNH R311 (partial skull).

Locality and Horizon

All material thus far referred to this species is from the Lower Lias of Lyme Regis, Dorset. The geological range is taken to be from the zone of *Schlotheimia angulata* to that of *Arnioceras semicostatum* (see Fig. 5) [Upper Hettangian to Lower Sinemurian].

Description

The present species is known only from cranial material. The holotype, a well preserved laterally compressed skull, exposed on the right, has a jaw length of 70 cm, and is the smallest of the three referred specimens; BMNH R311, the largest, has a jaw length of 83 cm. Characteristic of the species is the recurved snout which imparts the sardonic expression reflected in the trivial name. Ratio, length of snout to length of jaw 0.59 in holotype, observed range 0.59–0.62 (n = 3), (0.62–0.64 in *T. platyodon*, n = 3). Orbit relatively large, ratio, diameter of orbit to length of jaw 0.20 in the holotype, observed range 0.19–0.20 (n= 3), (0.16–0.17 in *T. platyodon*, n = 3). Sclerotic ring well preserved only in holotype; ratio, internal diameter of sclerotic ring to diameter of orbit 0.26 (0.24–0.31 in *T. platyodon*, n = 4). Relative small size of maxilla reflected in large value of ratio, length of premaxillary segment to length of jaw, 0.39 in holotype, range 0.38–0.39 (n = 3), (0.30–0.34 in *T. platyodon*, n = 2). Because of distortion external naris in holotype difficult to delimit; in BMNH R311 it is relatively small, ratio, length of external naris to diameter of orbit 0.44 (0.53–0.72 in *T. platyodon*, n = 4). Ratio, length of prenarial segment to length of jaw 0.44 in holotype, observed range 0.44–0.49 (n = 3), (0.46–0.50 in *T. platyodon*, n = 3).

Remarks

T. risor is most closely related to *T. platyodon*, but is a much smaller species (observed size ranges, based on jaw length, 70–83 cm and 135–

221 cm respectively). In marked contrast to *T. platyodon*, the snout of *T. risor* has a curved profile and also appears to be relatively shorter. In possessing a relatively larger orbit and smaller maxilla than *T. platyodon*, *T. risor* represents a more advanced stage of cranial evolution (for a discussion of evolutionary trends see McGowan, 1972a and 1972b).

The superficial resemblance to *I. communis* is due to their similar snout proportions; the mean value for the ratio, length of snout to length of jaw is 0.61 in *T. risor* and 0.63 in *I. communis*. The differences between the two species are fundamental:

	T. risor	*I. communis*
Diameter of orbit/length of jaw	< 0.21	> 0.21
Internal diameter of sclerotic ring/diameter of orbit	< 0.35	> 0.35
Length of premaxillary segment/length of jaw	< 0.40	> 0.40

Furthermore, *T. risor* is a considerably larger species, having an observed size range of 70–83 cm (jaw length) whereas the largest recorded specimen of *I. communis* has a jaw length of only 54 cm.

Temnodontosaurus eurycephalus sp. nov.

Fig. 9

Etymology

ευρυσ Gk. (masc.) wide; κεφαλη Gk. (fem.) head

Ichthyosaurus breviceps Owen (*pro-parte*), 1881, p. 109, pl. 29, fig. 1.
[*non*] *Ichthyosaurus platyodon* Conybeare, Lydekker, 1889a, p. 98.
Eurypterygius breviceps (Owen), Von Huene, 1922, p. 8, pl. 1, fig. 3 [referred to in the figure legend, in error, as *Eurypterygius brevirostris* (Owen)].

Diagnosis

Large ichthyosaurs, adult jaw length exceeding 100 cm; snout relatively short, ratio, length of snout to length of jaw < 0.58; skull and lower jaw deep; orbit relatively small, ratio, diameter of orbit to length of jaw < 0.21; ratio, internal diameter of sclerotic ring to diameter of orbit < 0.35; maxilla relatively long, ratio, length of premaxillary segment to length of jaw < 0.36 and probably < 0.30; external naris relatively short, ratio, length of external naris to diameter of orbit <0.50; ratio, length of prenarial segment to length of jaw < 0.42; teeth relatively few in number, maxillary tooth count probably <15. Lower Liassic in age [Upper Hettangian to Lower Sinemurian] and largely if not wholly confined to Lyme Regis, Dorsetshire.

Holotype

BMNH R1157, an isolated skull in the British Museum (Natural History).

Other Material

Nil

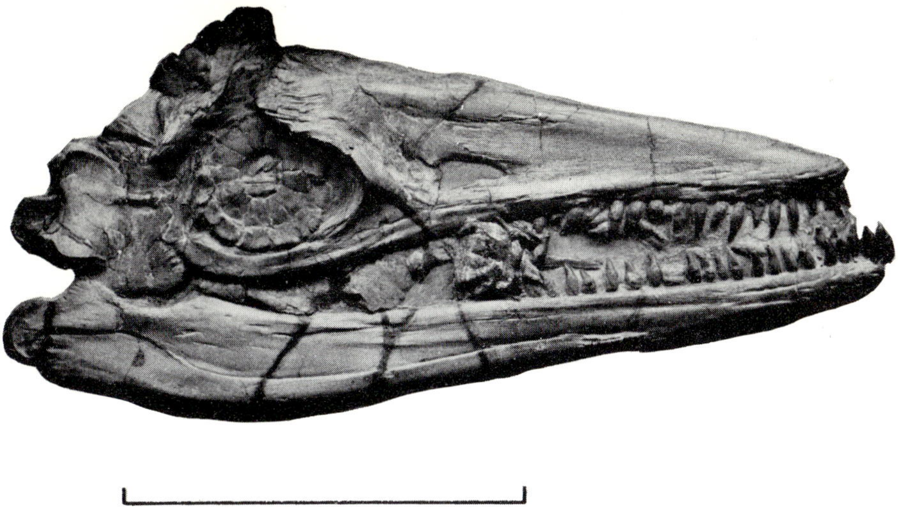

Fig. 9 *Temnodontosaurus eurycephalus* sp. nov., BMNH R1157, holotype, Lower Lias, Lyme Regis, scale = 50 cm.

Locality and Horizon

The holotype, from the Lower Lias of Lyme Regis, was collected from a limestone bed called Broad Ledge. Broad Ledge is probably synonymous with Grey Ledge (Woodward and Ussher, 1911, p. 37), being the top of a platform of rocks exposed at low tide some distance from the eastern jetty, southeast of Church Cliffs. The horizon corresponds with the zone of *Arietites bucklandi* and is lowermost Sinemurian (see Fig. 5).

Description

The holotype is a large laterally compressed skull, right side exposed. Length of lower jaw 111 cm, skull 102 cm. Snout much abbreviated, skull roof slopes quite steeply towards tip of snout. Ratio, length of snout to length of jaw 0.55 (0.62–0.64 in *T. platyodon*, n = 3). Skull and lower jaw both very deep dorsoventrally, giving head a robust appearance. Orbit and aperture of sclerotic ring both relatively small; ratio, diameter of orbit to length of jaw 0.18 (0.16–0.17 in *T. platyodon*, n = 3), ratio, internal diameter of sclerotic ring to diameter of orbit 0.29 (0.24–0.31 in *T. platyodon*, n = 4). Maxilla long and fairly stout, ratio, length of premaxillary segment to length of jaw only 0.27 (0.30–0.34 in *T. platyodon*, n = 2). External naris set well forward and relatively small; ratio, length of external naris to diameter of orbit 0.44 (0.53–0.72 in *T. platyodon*, n = 4); length of prenarial segment to length of jaw only 0.38 (0.46–0.50 in *T. platyodon*, n = 3).

Dentition

Teeth, large conical pegs with bulbous roots, few in number, and well spaced. The stout teeth, massive lower jaw, and broad rostrum are all indicative of an effective crushing apparatus. *T. eurycephalus* may have preyed upon other ichthyosaurs: the bony element (basisphenoid) clenched between the teeth of BMNH R1157 may not be part of its own skull but that of some victim.

Remarks

BMNH R1157 was figured and described by Owen in 1881 and referred to the new species *Ichthyosaurus breviceps*, even though there was a considerable size disparity between this specimen (jaw length 111 cm) and the holotype of *I. breviceps*, BMNH 43006 (jaw length 25.3 cm). The only similarity between the two is in the extreme brevity of the snout; the ratio, length of snout to length of jaw is 0.55 in BMNH R1157 and in BMNH 43006. However, the orbit is much smaller in BMNH R1157, and the ratio, diameter of orbit to length of jaw is only 0.18 compared with 0.31 in BMNH 43006. This discrepancy cannot be reconciled in terms of relative growth because the growth of the orbit appears to have had a positive allometry in *I. breviceps* (McGowan, 1973a). The maxilla is relatively longer in BMNH R1157 than in BMNH 43006, with values for the ratio, length of premaxillary segment to length of jaw 0.27 and 0.44 respectively. Here again the disparity cannot be explained in terms of relative growth because the maxilla appears to have had a negative allometry in *I. breviceps*, larger individuals tending to have relatively smaller maxillae (McGowan, 1973a). Possession of small orbit, small sclerotic ring diameter and large maxilla are all longipinnate characters (McGowan, 1972a), and there can be no doubt that BMNH R1157 is a longipinnate. BMNH R1157 cannot therefore be referred to the latipinnate species *I. breviceps*, and is accordingly described here as a new longipinnate species.

Temnodontosaurus longirostris (Mantell)

Fig. 10

Ichthyosaurus longirostris Mantell, 1851, p. 385.
[*non*] *Ichthyosaurus longirostris* Mantell, Jäger, 1856, p. 951, pl. 30.
Ichthyosaurus longirostris Mantell, Owen, 1881, p. 124, pl. 25, fig. 2; pl. 28, fig. 3; pl. 32, figs. 7, 8, 9.
Ichthyosaurus latifrons Koenig, Lydekker, 1889a, pp. 89-92, figs. 31-32.
Ichthyosaurus longirostris Mantell, Fraas, 1891, p. 63.
Leptopterygius laitfrons (Koenig), Von Huene (*pro-parte*), 1922, p. 15, pl. 3, fig. 2.

Emended Diagnosis

An inadequately known species of moderate size, length of jaw probably not exceeding 100 cm; snout very long and thin, ratio, length of snout to length of jaw > 0.70; orbit relatively small, ratio, diameter of orbit to length of jaw probably $\leqslant 0.18$; sclerotic aperture relatively large for a longipinnate, but ratio, internal diameter of sclerotic ring to diameter of orbit probably always < 0.35; forefin long and narrow with three primary digits and one post-axial accessory digit; radius notched, perhaps also

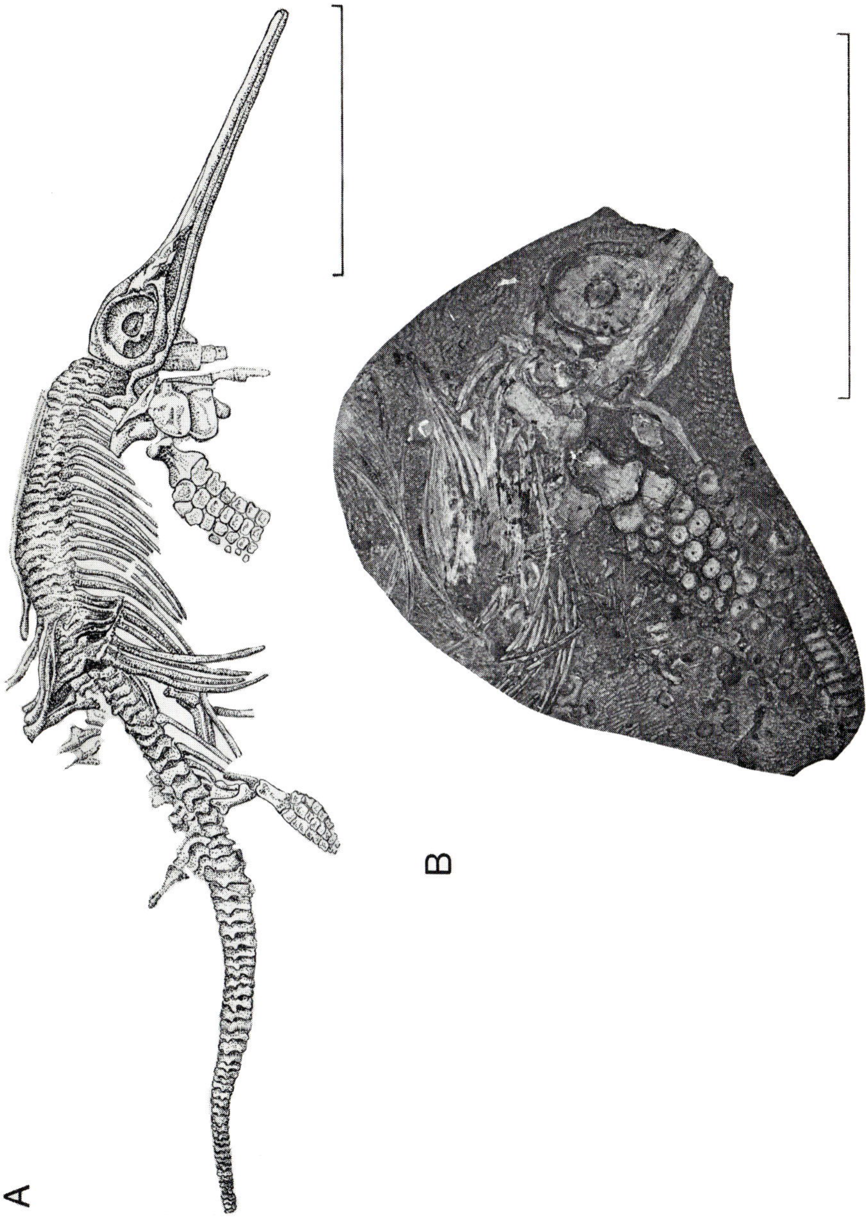

Fig. 10 *Temnodontosaurus longirostris* (Mantell), scale = 50 cm.
A. BMNH 36182, Lower Lias, Barrow-on-Soar, Leicestershire, after Owen, 1881, pl. 32, fig. 7; B. SCC 1, Lower Lias, Street, Somersetshire.

radiale and other elements in first digit; maximum number of elements in longest digit probably < 18. Liassic in age, probably ranging from Lower to Upper divisions.

Holotype
BMNH 14566, an incomplete skeleton (dorsal aspect exposed) apparently from the Upper Lias of Whitby, Yorkshire, figured in Owen, 1881, pl. 32, fig. 8.

Referred Material
BMNH 36182, an almost complete skeleton from the Lower Lias of Barrow-on-Soar; SCC 1, the anterior portion of an imperfect skeleton comprising partial skull, near-complete forefin, and an imperfect pectoral girdle, from the Lower Lias of Street, Somerset. BMNH R1120, a poorly preserved skeleton from the Lower Lias of Lyme Regis figured by Owen (1881, pl. 32, fig. 1) is also referred to the present species.

Locality and Horizon
According to Lydekker (1889a, p. 91) the holotype is from the Upper Lias of Whitby, Yorkshire, whereas the two referred specimens, BMNH 36182 and SCC 1 are both from the Lower Lias of Barrow-on-Soar, Leicestershire, and Street, Somerset, respectively. As far as I am aware, no Liassic species extends from Lower to Upper divisions, but this may be an artifact of the geological record, due to lack of good Upper Liassic exposures (ichthyosaurs with very extensive geological ranges are known from the Cretaceous, McGowan, 1972c). It could be that Lydekker was mistaken in referring BMNH 14566 to the Upper Lias of Whitby; the Lower Lias does outcrop in Whitby, but most if not all ichthyosaurs from Whitby have been collected from the Upper Lias. For the present the geological range of *T. longirostris* is taken to extend from Lower to Upper Lias.

Most of the reptilian remains from Street have been collected from the "pre-*planorbis* Beds" (Arkell, 1933, p. 123) which are at the base of the Lower Lias [Lowermost Hettangian]. At Whitby ichthyosaurs have been collected from the zones of *Hildoceras bifrons* and *Harpoceras falcifer* (Hemingway *et al.*, 1968) which are in the lower part of the Upper Lias [Lower Toarcian]. The geological range of *T. longirostris* is therefore taken to be from Lower Hettangian to Lower Toarcian.

Emended Description
Because of incomplete material, *Temnodontosaurus longirostris* (Mantell) is only moderately well known.

SKULL
Snout long and thin, ratio, length of snout to length of jaw approximately 0.79 in holotype, compared with 0.72–0.73 (n = 2) in the other English long-snouted species *Ichthyosaurus tenuirostris* (0.62—0.64 in *T. platyodon*, n = 3). Orbit appears large but ratio, diameter of orbit to length of jaw only 0.16 in holotype and BMNH R1120, compared with 0.22 in *I. tenuirostris* (n = 1) and 0.16–0.17 in *T. platyodon* (n = 3). Sclerotic

aperture also relatively small, ratio, internal diameter of sclerotic ring to diameter of orbit 0.33 in holotype and SCC 1, compared with 0.41 in *I. tenuirostris* (n = 1) and 0.24–0.31 in *T. platyodon* (n = 4).

FINS

Knowledge of the forefin is incomplete, but it is certainly longipinnate, with three primary digits and one post-axial accessory digit (Fig. 10). The best known forefin (SCC 1) is almost complete and has a maximum number of 16 elements in the longest digit, though one or two may be missing from the extreme tip. Only the radius is notched in SCC 1 whereas in BMNH 36182 and BMNH R1120 the radius, radiale, and third element of the first digit are emarginated. The hindfin of BMNH 36182 has three primary digits and several of the elements of the first digit are emarginated.

Remarks

The species *I. longirostris* is usually attributed to Owen, 1881 (Lydekker, 1889a, p. 89; Woodward and Sherborn, 1890, p. 239; Kuhn, 1934, p. 55) and sometimes to Jäger 1856 (Lydekker, 1889a, p. 89) but the first description of the species was given by Mantell in 1851, who wrote, "ICHTHYOSAURUS LONGIROSTRIS Wall-case E—In . . . Case E there is part of the skeleton of an Ichthyosaurus from Whitby, about six feet [1.8 m] in length. It is remarkable for the exceedingly slender and elongated muzzle; the skull is crushed; and with the exception of the chain of vertebrae which extends to the tail, and a few bones of one paddle, there are no characteristic parts preserved. The specific name, longirostris, is affixed to this specimen; but I cannot ascertain that it is figured or described." (Mantell, 1851, p. 385). Sufficient information is given to identify the specimen, the locality is recorded and, most important, note is taken of the extremely long snout. In my opinion this constitutes a valid description. Lydekker (1889a, p. 91) identified Mantell's specimen as BMNH 14566, from the Upper Lias of Whitby, and gave a description which matches Mantell's description, noting that the dorsal aspect of the skull is figured by Owen (1881, pl. 32, fig. 8). Jäger (1856, p. 951) did not consider that Mantell had adequately described the species and therefore gave his own description which included the figure of a skull (Jäger, 1856, pl. 30). This skull, however, should probably be referred to the genus *Eurhinosaurus*. The skeleton figured by Fraas (1891, pl. 12, fig. 5) from the Upper Lias of Holzmaden might be referable to *T. longirostris*.

Systematic Summary

The genus *Temnodontosaurus*, which is largely confined to the Lower Lias, contains four species, *T. platyodon* (Conybeare), *T. longirostris* (Mantell), and the two new species *T. risor* and *T. eurycephalus*.

Still to be considered is *I. acutirostris Owen*, from the Upper Lias of Whitby, Yorkshire. *I. acutirostris* and *Stenopterygius quadriscissus* (Quenstedt) [Upper Lias, Germany] are found to be almost indistinguishable from each other but quite distinct from *T. platyodon* (Table 1). *I. acutirostris* is accordingly referred to the genus *Stenopterygius*. Because the present work is confined to English material it is not appropriate to give here an emended diagnosis of the family Stenopterygiidae, which is known largely from Germany. This will be the subject of another study (McGowan, in preparation).

Table 1. Comparison of cranial and body characters between the longipinnate species *S. acutirostris*, *S. quadriscissus* and *T. platyodon*.

	S. acutirostris	*S. quadriscissus*	*T. platyodon*
*Length of snout/length of jaw	0.67	0.67	0.63
*Diameter of orbit/length of jaw	0.20	0.20	0.16
*Length of premaxillary segment/ length of jaw	0.46	0.47	0.34
*Internal diameter of sclerotic ring/ diameter of orbit	0.34	0.38	0.28
*Length of prenarial segment/ length of jaw	0.57	0.56	0.49
*Length of external naris/ diameter of orbit	0.35	0.39	0.66
Number of primary digits	3	3	3
Number of accessory digits	1 or 2	2	1
Maximum number of elements in longest digit	16	17	14
Maximum recorded size (jaw length, cm)	153	55	221

*Mean values

Family Stenopterygiidae Von Huene 1948
Genus *Stenopterygius* Jaekel 1904

Stenopterygius acutirostris (Owen)
Fig. 11

Ichthyosaurus acutirostris Owen, 1840, p. 121.
Ichthyosaurus longipennis Mantell, 1851, p. 378.
Ichthyosaurus acutirostris Owen, Phillips, 1875, p. 272.
Ichthyosaurus acutirostris Owen, Blake, 1876, p. 253.
Ichthyosaurus crassimanus Blake, 1876, p. 253, pl. 1, fig. 9.
Ichthyosaurus zetlandicus Seeley, 1880, pp. 635-646, pl. 25.
Ichthyosaurus acutirostris Owen, Owen, 1881, p. 121, pl. 28, fig. 2.
Ichthyosaurus longifrons Owen, 1881, p. 118, pl. 23, figs. 1-5; pl. 24, fig. 1; pl. 25, fig. 1; pl. 26, fig. 1; pl. 27, figs. 2-5.
Ichthyosaurus acutirostris Owen, Lydekker, 1889a, p. 73.
Leptopterygius acutirostris Owen, Von Huene (*pro-parte*), 1922, p. 25.
[*non*] *Ichthyosaurus* (*cf.*) *acutirostris* Owen, Lydekker, 1889a, p. 74, fig. 27.
[*non*] *Ichthyosaurus acutirostris* Owen, Fraas, 1891, pl. 8; pl. 11, fig. 1.
[*non*] *Leptopterygius acutirostris* (Owen), Hauff, 1953, pl. 19.

Emended Diagnosis

Large ichthyosaurs, length of jaw $>$ 60 cm in mature individuals, frequently $>$ 100 cm; snout relatively long, ratio, length of snout to length of jaw $\geqslant$ 0.62; snout tends to taper to a sharp point; orbit often relatively large, ratio, diameter of orbit to length of jaw usually not less than 0.18 and may exceed 0.20; sclerotic aperture relatively large, ratio, internal diameter of sclerotic ring to diameter of orbit frequently $>$ 0.35; maxilla relatively short, ratio, length of premaxillary segment to length of jaw $>$ 0.36; ratio, length of external naris to diameter of orbit $<$ 0.60; forefin with three primary digits and one or two accessory digits; head large relative to body, especially in larger individuals. Jaw and body length data approximate more closely to equation, jaw length $= 0.1 \times (\text{body length})^{1.2}$ than to the equation, jaw length $= 3.363 \times (\text{body length})^{0.507}$ (see remarks below). Upper Liassic in age, apparently confined to England, probably to the Whitby locality of Yorkshire.

Holotype

BMNH 14553 (Lydekker, 1889a, p. 73). There is some doubt regarding the present whereabouts of this specimen (Walker, *pers. comm.*), and only an extensive search of the British Museum would confirm whether it has been lost. At present this is not practical and it will therefore be recorded that the holotype may be lost.

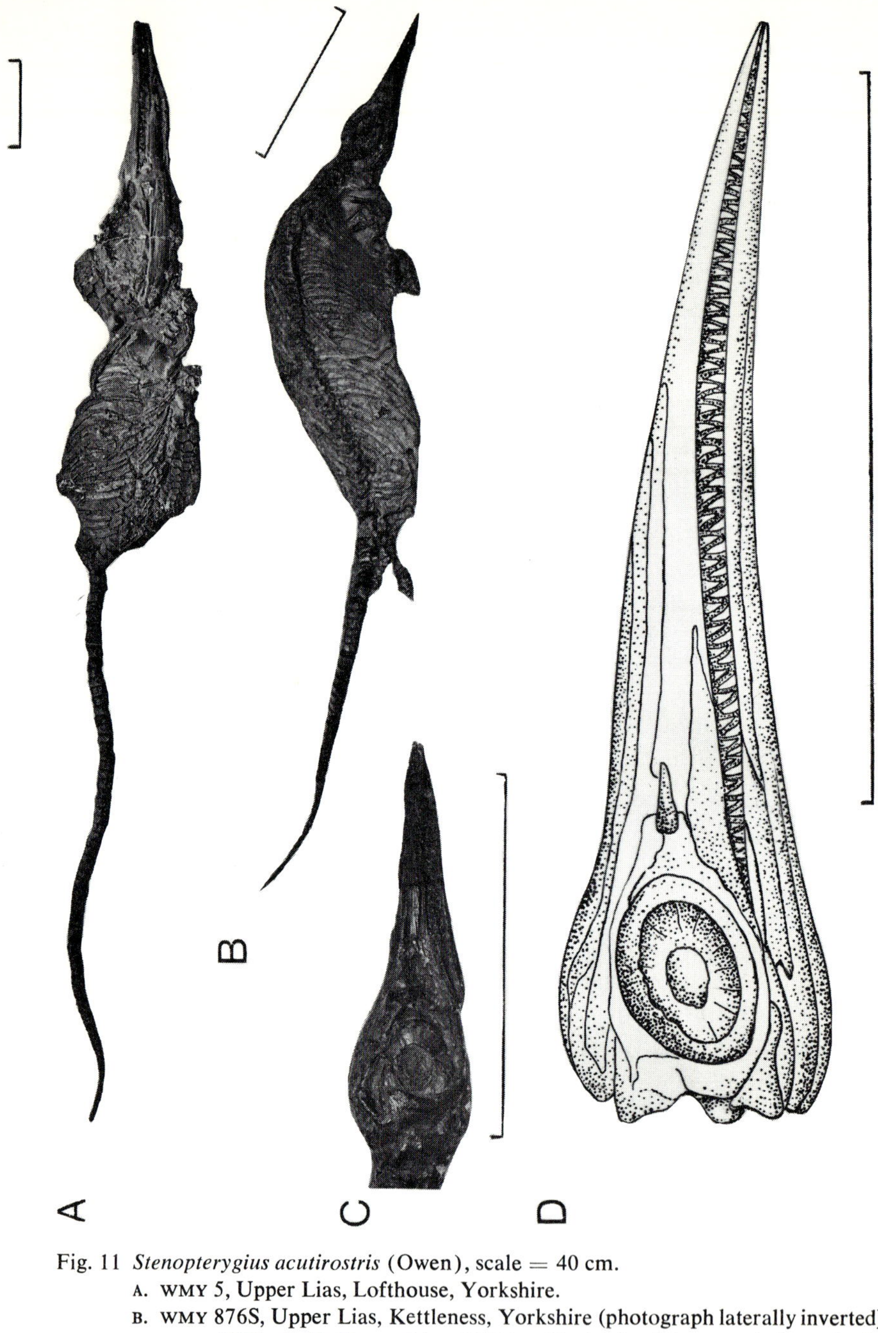

Fig. 11 *Stenopterygius acutirostris* (Owen), scale = 40 cm.

A. WMY 5, Upper Lias, Lofthouse, Yorkshire.

B. WMY 876S, Upper Lias, Kettleness, Yorkshire (photograph laterally inverted).

C. WMY 877S, skull, Upper Lias, Whitby, Yorkshire (photograph laterally inverted).

D. BMNH 14533, skull of holotype, Upper Lias, Whitby, Yorkshire (after Owen, 1881, pl. 28, fig. 2).

Other Material

WMY 5, WMY 2546S, WMY 876S, WMY 878S, YM 497 (complete or near complete skeletons); WMY 877S (skull with some post-cranials); WMY 876S(H1), BMNH 1500a (complete skulls) and SMC J35176 (partial skull).

Locality and Horizon

All the material referred to this species is from the Upper Lias, in the vicinity of Whitby, Yorkshire. Most of the specimens were collected from the Alum Shales, and also from the Hard Shales and Bituminous Shales lying above the Jet Rock (see Arkell, 1933, p. 182; Wilson, 1948, p. 28) in the ammonite zone of *Hildoceras bifrons* and *Dactylioceras commune* [Lower Toarcian].

Emended Description

Because of poor preservation the species is not adequately known, even though it is represented by at least ten specimens. The holotype is apparently not well preserved.

SKULL

The type skull, figured by Owen (1881, pl. 28, fig. 2) is reproduced in Fig. 11D (Lydekker, 1889a, p. 74, erroneously refers to Owen's figure as pl. 32, fig. 8). Snout frequently sharply pointed, as in WMY 876S, BMNH 1500a, WMY 876S(H1) and WMY 878S, but decurved appearance figured by Owen, seen clearly in WMY 876S, probably an artifact of preservation because it is not a constant feature. Decurved snouts in BMNH 1500a and WMY 876S(H1) almost certainly due to dorso-ventral compression, the thin snout being readily distorted. Snout quite straight in WMY 876S and WMY 877S, even slightly recurved in WMY 5. Ratio, length of snout to length of jaw 0.62–0.68 (n = 6), mean value 0.66 (0.62–0.64 in *T. platyodon*, n = 3). Measurements taken from Owen's figure of type skull give a value of 0.78 for this ratio, but this may be because the snout was exaggerated in his figure.

Maxilla relatively shorter than in the *Temnodontosaurus*, indicating a more advanced stage of cranial evolution (McGowan, 1972b); ratio, length of premaxillary segment to length of jaw 0.40–0.42 (n = 2), mean 0.41 (0.30–0.34 in *T. platyodon*, mean 0.32, n = 2). Orbital condition similarly more advanced than *Temnodontosaurus*, ratio, diameter of orbit to length of jaw 0.16–0.23 (n = 4), mean 0.20 (0.16–0.17 in *T. platyodon*, mean 0.17, n = 3). Sclerotic aperture variable, ratio, internal diameter of sclerotic ring to diameter of orbit 0.28–0.40 (n = 4) compared with 0.24–0.31 (n = 4) in *T. platyodon*. Ratio, length of external naris to diameter of orbit (not always a reliable character) relatively small, 0.31–0.57 (n = 4), mean 0.42, compared with 0.53–0.72 mean 0.63 (n = 4) in *T. platyodon*. Internasal foramen absent (because of poor preservation this can only be determined in one specimen, SMC J35176, the holotype of Seeley's *I. zetlandicus*).

FINS

Fins generally poorly preserved, but are determinate in YM 497, WMY 5, WMY 877S, WMY 2546S and apparently in holotype. Three primary and one accessory digits, except in WMY 2546S, which also has one pre-axial accessory digit. Maximum number of elements in longest digit an estimated 14 in WMY 5, 16 in WMY 2546S and 19 or 20 in YM 497. Notching in radius, radiale and third element in WMY 5, in radius and radiale in WMY 2546S, only in radius in YM 497, with no notching in WMY 877S. Forefin not as long and narrow as in *T. platyodon*, aspect ratio only 2.6 (n = 3) compared with 3.1 in *T. platyodon* and 2.3 in *S. quadriscissus* (mean values). Relative lengths of fore- and hindfins variable, ratio, length of forefin to length of hindfin 2.2 in YM 497, 1.2 in WMY 2546S (2.5 in *S. quadriscissus*, 1.4 in neotype of *T. platyodon*).

In WMY 2546S left fore- and hindfins are composites with individual elements set in plaster in a latipinnate configuration (four primary digits). Fins on the right side appear authentic and have a longipinnate configuration with three primary digits. Of interest in WMY 5 is that the humerus has three distal facets (comparable with *Platypterygius*, see McGowan, 1972c; and *Ophthalmosaurus*, see Andrews, 1910), the smaller of these probably being for the pisiform. It is not known whether this is a constant feature of the species. To summarize, the forefin has three primary digits with one post-axial accessory digit and sometimes one pre-axial accessory. Maximum number of elements in longest digit up to 20. Notching usually present.

AXIAL SKELETON AND GIRDLES

Vertebral count to pelvic girdle between 48 and 50 in WMY 2546S, and approximately 48 in WMY 876S. Vertebral count to tail bend difficult to determine; in WMY 2546S there is some indication of a disturbance at level 77 and another at level 93, either of which could indicate the position of the tail bend. In WMY 876S there is some indication of the bend at about level 76. It is tentatively concluded that the vertebral count to the tail bend is approximately 76.

Partial pectoral girdles preserved in WMY 878S and WMY 5 but comprise little more than the coracoids. The coracoids are fairly thick and robust and the pectoral girdle may have been well developed. Anterior notch present.

DENTITION

The teeth are of the usual ichthyosaurian form but appear to be relatively small (well marked in SMC J35176).

Remarks

1. Reasons for considering *S. acutirostris* to be distinct from *S. quadriscissus*

The only difference between *S. acutirostris* and *S. quadriscissus* is size, the former reaching about three times the size of the latter (based on jaw length), and it seemed reasonable to conclude that they were part of a continuous growth series of one and the same species. However, the larger individuals have relatively larger heads than the smaller ones, contrary to the usual ontogenetic pattern. Allometric growth constants have recently been evaluated for *S. quadriscissus* (McGowan, 1973a) and the relationship between the growth of the head relative to the body has been found to conform to the equation $y = 3.363x^{0.507}$ (from the allometric growth equation $y = bx^{\alpha}$ where y = organ size, x = body size, α = the allometric growth constant and b = a second constant). In *S. acutirostris*, head growth relative to that of the body conforms to the equation $y = 0.1x^{1.2}$ but the small sample size (n = 4) does not permit confidence limits to be set for the estimates of α and b.

Three possible conclusions may be drawn from this:

A. *S. quadriscissus* and *S. acutirostris* represent males and females of a sexually dimorphic species.

B. *S. quadriscissus* and *S. acutirostris* belong to a single species in which changes in allometric growth constants α and b occur during ontogeny.

C. *S. quadriscissus* and *S. acutirostris* are distinct species differing in allometric growth constants and maximum attainable sizes.

S. quadriscissus and *S. acutirostris* occur in separate localities (in Germany and England respectively), which may not be strictly contemporaneous. Sexual dimorphism may therefore be dismissed.

Abrupt changes in allometric growth constants are uncommon in ichthyosaurs (McGowan, 1973a), but occur in other animals, usually associated with metamorphosis. Martin (1949) found growth in fishes to be characterized by a series of five stanzas corresponding to four larval stages and the sexually mature adult. Each stanza had a different set of values for α and b. In *Rana temporaria*, the allometric growth constant of the eye undergoes an abrupt change at metamorphosis (De Jongh, 1967), and Brown and Davies (1972) working on the cockroach *Ectobius*, found the majority of changes in growth constants coincided with metamorphic changes.

If *S. quadriscissus* and *S. acutirostris* were part of a continuous growth series the abrupt change in allometric growth constants could not be correlated with sexual maturity because pregnant females occur in the smaller form (*S. quadriscissus*). Maternal specimens are relatively small; 6293 (Stuttgart Museum) has a body length of only 219 cm (See Fraas, 1891, p. 52, pl. 4, fig. 2) compared with 628 cm in the largest specimen of *S. acutirostris* (WMY 2546S). If the two species were part of the same growth series it would suggest attainment of sexual maturity at an early stage of growth.

Acquisition of more extensive data from Germany could possibly alter the position, but for the present it is concluded that *S. quadriscissus* and *S. acutirostris* are separate species.

2. Reasons for synonymizing *I. crassimanus*, *I. longifrons* and *I. zetlandicus* with *S. acutirostris*

Blake's (1876, p. 253) description of *I. crassimanus* was mainly based on a large specimen in the York Museum which was redescribed in greater detail by Melmore in 1930. The specimen (YM 497), collected from the Alum Shales north of Whitby, is unfortunately incomplete so that few measurements can be taken. The anterior portion of the snout is missing and is reconstructed in plaster, and the body is incomplete beyond the tail bend. If allowances are made for these missing parts, the head and body proportions approximate to the growth equation evaluated above for *S. acutirostris*. The forefin has three primary and one post-axial accessory digits, and is consistent with *S. acutirostris*. The only inconsistency is in the relative proportions of fore- and hindfins; in WMY 2546S (*S. acutirostris*) the ratios, forefin length to hindfin length and forefin width to hindfin width have values of 1.2 and 1.3, whereas in YM 497, the values are 2.2 and 1.6 respectively. However, the relative size of fore- and hindfin is a variable character; for example in a sample of 10 specimens of *S. quadriscissus* the ratio, forefin length to hindfin length has a range of 2.2–3.1. The aspect ratio of the forefin is 2.4 in WMY 2546S and 2.9 in YM 497. It is therefore concluded that there are no major inconsistencies between YM 497 and WMY 2546S and since YM 497 is from the same locality and horizon as *S. acutirostris* it is referred to that species. *I. crassimanus* Blake is thus a junior synonym for *S. acutirostris*.

In describing *I. zetlandicus* Seeley wrote, "I am acquainted with no other ichthyosaur in which the skull attains this broad, triangular form, with the orbits so far apart from each other and so moderately inclined, and with the nares so far in front of the orbits and relatively so large . . . These characters sufficiently distinguish the species from all others." (Seeley, 1880, p. 646). The holotype (SMC J35176), a large incomplete skull from the Upper Lias of Whitby, is preserved in-the-round and the absence of distortion gives a very broad appearance when viewed from above. However, compared with other in-the-round skulls (BMNH R8177, BMNH 49203, both *I. communis*), SMC J35176 is no broader nor are the orbits any further apart. Because the skull is incomplete few comparisons can be made, but the ratio, length of external naris to the diameter of orbit is larger (0.57) than it is in *S. acutirostris*. However, the sample size is small ($n = 3$), and this character is known to be variable (for example in *Stenopterygius quadriscissus* the observed range is 0.25–0.45, $n = 9$). Furthermore the holotype of *I. zetlandicus* is from the same locality and horizon as *S. acutirostris*. *I. zetlandicus* is therefore concluded to be a junior synonym for *S. acutirostris*.

Owen's description of *I. longifrons* comprised a brief reference to some figures (1881, pls. 23-27). The skull figured is identified by Lydekker as BMNH 33157, from the Upper Lias of Curcy (1889a, p. 78). Lydekker noted that, "In all essential characters this specimen agrees with the preceding [a cast of the holotype of *I. zetlandicus*] . . ." Lydekker synonymised *I. longifrons* with (the now invalid) *I. zetlandicus*, a course taken by most authors (see Woodward and Sherborn, 1890). Only one character can be measured in BMNH 33157, the ratio, length of external naris to diameter of orbit (0.38) which lies within the observed ranges for *S. acutirostris* (0.31–0.57). *I. longifrons* is thus concluded to be a junior synonym of *S. acutirostris*.

3. Reasons for rejecting *I. latifrons* Koenig as a *nomen dubium*

I. latifrons was first mentioned by Koenig who figured an incomplete skull and partial vertebral column (1825, pl. 19, fig. 250) and most authorities consider this sufficient to make the name available (Woodward and Sherborn, 1890; Lydekker, 1889a, Kuhn, 1934). The figure was not accompanied by a written description, and the specimen (BMNH R1122) is quite indeterminate. Owen figured the skull (1881, pl. 27, fig. 1) and Lydekker noted that it was probably collected from the Lower Lias of Barrow-on-Soar, Leicestershire (1889a, p. 90, fig. 33). Because the holotype is indeterminate it is not possible to compare it with any other material. The name *I. latifrons* is thus not applicable to any known taxon and is accordingly rejected as a *nomen dubium*.

Conclusion

Of the five species of longipinnate ichthyosaurs of the English Lower Jurassic three are from the Lower Lias: *Temnodontosaurus platyodon* (Conybeare), *T. risor*, sp. nov., and *T. eurycephalus*, sp. nov. *T. longirostris* (Mantell) extends from Lower to Upper Lias, while *Stenopterygius acutirostris* (Owen) is confined to the Upper Lias. *T. platyodon* and *S. acutirostris* are both large, the former reaching a total length of 9 m, the latter approaching 8 m. Body length cannot be estimated for *T. eurycephalus*, but as the skull is in excess of 1 m, the species is assumed to have been comparable in size to *T. platyodon* and *S. acutirostris. T. risor* and *T. longirostris* were probably both of more moderate size, and much smaller than either *T. platyodon* or *S. acutirostris.*

T. eurycephalus, with its short snout and massive skull, and *T. longirostris*, with its long and delicate snout were the most highly specialized species. It was suggested above that *T. eurycephalus* may have fed upon other ichthyosaurs, and the short deep skull and heavy teeth are certainly suggestive of a voracious predator. On the other hand, *T. longirostris*, with its long thin snout, probably fed exclusively on small fishes.

S. acutirostris differs from its German contemporary *S. quadriscissus* in being larger and having a relatively larger head. Indeed the two species are so similar that they might well have evolved from similar stocks by way of a change in the allometric growth constants of the head and body.

Acknowledgments

I wish to express my sincere thanks for the help and hospitality I received from the staffs of the many institutions visited. I particularly wish to thank Dr. Alan Charig, Mr. Cyril Walker, Mr. Phillip Palmer, and Miss A. L. Holloway of the British Museum (Natural History); Dr. Colin Forbes of the Sedgwick Museum and Mr. Peter Biddlestone, formerly of the Sedgwick Museum, Cambridge, now of the Royal Ontario Museum; Mr. Phillip Powell of the Oxford University Museum; Mr. Allen Butterworth and Ms. Barbara Pyrah of the Yorkshire Museum; Mr. J. G. Graham of the Whitby Museum, Yorkshire; Dr. J. E. Hemingway of the University of Newcastle-upon-Tyne; Dr. Paul McGrew of the University of Wyoming; Mr. Andrew Mathieson and Mr. M. J. Jones of the Leicester County Museum, and the staff of the Clarks Shoe Museum, Street, Somerset.

I thank Mrs. Janet Clarke and Mrs. Lynda Spicer, ROM, for their secretarial help; Mrs. Sophie Poray, ROM, for the drawings; Mr. Leighton Warren and Mr. Alan McColl, Photography Department, ROM, for photographs; and Miss E. Dowie of the ROM library for assistance with literature searches. For reading the manuscript and making many helpful suggestions I thank Dr. Allan Baker, Department of Ornithology, ROM and Dr. Gordon Edmund, Department of Palaeontology, ROM.

Literature Cited

ANDREWS, C. W.

1910 A descriptive catalogue of the marine reptiles of the Oxford Clay based on the Leeds Collection in the British Museum (Natural History), London. Part 1. London, Printed by order of the Trustees of the British Museum. 205 pp.

ARKELL, W. J.

1933 The Jurassic system in Great Britain. Oxford, Clarendon Press. 681 pp.

BLAKE. J. F.

1876 Order Ichthyopterygia. *In* Tate, R. and J. F. Blake. The Yorkshire Lias. London, J. Van Voorst, pp. 246-261.

BROWN, V. AND R. G. DAVIES

1972 Allometric growth in two species of *Ectobius* (Dictyoptera: Blattidae). Proc. Zool., vol. 166, pt. 1, pp. 97-132.

CONYBEARE, W. D.

1822 Additional notices on the fossil genera *Ichthyosaurus* and *Plesiosaurus*. Trans. Geol. Soc. Lond., ser. 2, vol. 1, no. 1, pp. 103-123.

DE JONGH, H. J.

1967 Relative growth of the eye in larval and metamorphosing *Rana temporaria*. Growth, vol. 31, pp. 93-103.

DEAN, W. T., D. T. DONOVAN AND M. K. HOWARTH

1961 The Liassic ammonite zones and subzones of the North-Western European Province. Bull. Br. Mus. Nat. Hist., Geol., vol. 4, no. 10, pp. 435-505.

DE LA BECHE, H. T.

1822 Remarks on the geology of the south coast of England, from Bridport Harbour, Dorset, to Babbacome Bay, Devon. Trans. Geol. Soc. Lond., ser. 2, vol. 1, pp. 40-47.

1826 On the Lias of the coast, in the vicinity of Lyme Regis, Dorset. Trans. Geol. Soc. Lond., ser. 2, vol. 2, no. 1, pp. 21-30.

DE LA BECHE, H. T. AND W. D. CONYBEARE

1821 Notice of the discovery of a new fossil animal, forming a link between the *Ichthyosaurus* and the crocodile, together with general remarks on the osteology of the *Ichthyosaurus*. Trans. Geol. Soc. Lond., ser. 1, vol. 5, no. 2, pp. 559-594.

DELAIR, J. B.

1960 The Mesozoic reptiles of Dorset (Part III: conclusion). Proc. Dorset Nat. Hist. Archaeol. Soc., vol. 81, pp. 59-85.

1969 A history of the early discoveries of Liassic ichthyosaurs in Dorset and Somerset (1779-1835) and the first record of the occurrences of Ichthyosaurs in the Purbeck. Proc. Dorset Nat. Hist. Archaeol. Soc., vol. 90, pp. 115-132.

FRAAS, O.

1891 Die Ichthyosaurier der Suddeutschen Trias- und Jura-Ablagerungen. Tübingen. 81 pp.

HAUFF, B.

1953 Das Holzmadenbuch, Ohringen, Verlag der Hohenlohe'schen Buchlandlung. 54 pp.

HAWKINS, T.

1834 Memoirs of Ichthyosauri and Plesiosauri, extinct monsters of the ancient earth. London, Rolfe and Fletcher. 58 pp.

1840 The book of the great sea-dragons, Ichthyosauri and Plesiosauri, . . . gedolim taninim, of Moses. Extinct monsters of the ancient earth. London, W. Pickering. 27 pp.

HEMINGWAY, J. E., V. WILSON AND C. W. WRIGHT

1968 Geology of the Yorkshire Coast. Geologists' Association guides no. 34, Rev. ed. Colchester, Benham. 47 pp.

HUENE, F. VON

1922 Die Ichthyosaurier des Lias und ihre Zusammenhänge. Monogrn. Geol. Paläont., ser. 1, heft 1, pp. 1-114.

1948 Short review of the lower tetrapods. *In* Du Toit A.L., ed. Robert Broom commemorative volume. Royal Society of South Africa special publication. Capetown, Published by the Society, pp. 65-106.

JAEKEL, O.

1904 Eine neue Darstellung von *Ichthyosaurus*. Z. Dt. Geol. Ges., vol. 56, pp. 26-34.

JÄGER, G.

1856 Uber eine neue Species von Ichthyosauren (*Ichthyosaurus longirostris* Owen et Jäger). Nebst Bemerkungen uber die ubrigen in der Liasformation Wurttembergs auf gefunden Reptilien. Nova Acta Acad. Caesar. Carol., vol. 25, pt. 2, pp. 939-967.

KOENIG, E. C.

1825 Icones Fossilum Sectiles. Centuria Prima. London. 4 pp.

KUHN, O.

1934 Ichthyosauria. *In* Quenstedt, W., ed. Fossilium Catalogus. 1: Animalia. Pars 63. Berlin, Junk, pp. 1-75.

LYDEKKER, R.

1889a Catalogue of the fossil Reptilia and Amphibia in the British Museum (Natural History). Part II. Containing the orders Ichthyopterygia and Sauropterygia. London, Printed by Order of the Trustees of the British Museum. 307 pp.

1889b Palaeozoology; Vertebrata. *In* Nicholson, H. A. and R. Lydekker. A manual of palaeontology for the use of students with a general introduction on the principles of palaeontology. 3d. ed. vol. 2, part 3. Edinburgh, W. Blackwood, pp. 889-1474.

MANTELL, G. A.

1851 Petrifactions and their teachings; or, a hand-book to the gallery of organic remains of the British Museum. London, H. G. Bohn. 496 pp.

MARTIN, W. R.

1949 The mechanics of environmental control of body form in fishes. Univ. Toronto Stud. Biol. Ser., no. 58, pp. 1-91.

MCGOWAN, C.

1969 The cranial morphology and interrelationships of the Lower Liassic ichthyosaurs. Ph.D. thesis, University of London. 498 pp.

1972a The distinction between latipinnate and longipinnate ichthyosaurs. Life Sci. Occ. Pap., R. Ont. Mus., no. 20, pp. 1-12.

1972b Evolutionary trends in longipinnate ichthyosaurs with particular reference to the skull and fore fin. Life Sci. Contr., R. Ont. Mus., no. 83, pp. 1-38.

1972c The systematics of Cretaceous ichthyosaurs with particular reference to the material from North America. Contr. Geol. vol. 11, no. 1, pp. 9-29.

1973a Differential growth in three ichthyosaurs: *Ichthyosaurus communis*, *I. breviceps* and *Stenopterygius quadriscissus* (Reptilia, Ichthyosauria). Life Sci. Contr., R. Ont. Mus., no. 93, pp. 1-24.

1973b The cranial morphology of the Lower Liassic latipinnate ichthyosaurs of England. Bull. Br. Mus. Nat. Hist., Geol., vol. 24, no. 1, pp. 1-109.

In Press A revision of the latipinnate ichthyosaurs of the Lower Jurassic of England (Reptilia: Ichthyosauria). Life Sci. Contr., R. Ont. Mus.

MELMORE, S.

1930 A description of the type-specimen of *Ichthyosaurus crassimanus*, Blake (Owen MS). Ann. Mag. Nat. Hist., ser. 10, vol. 6, pp. 615-619.

OWEN, R.

1840 Report on the British fossil reptiles. Part 1. Rep. Br. Ass. Advmt. Sci., vol. 9, pp. 43-126.

1881 A monograph of the fossil Reptilia of the Liassic formations. Part III. *Plesiosaurus*, *Dimorphodon*, and *Ichthyosaurus*. [Part III] Order Ichthyopterygia . . . *Ichthyosaurus*. *Palaeontogr. Soc.* [Monog.] 35, pp. 83-134.

PHILLIPS, J.

1875 Illustrations of the geology of Yorkshire; or, a description of the strata and organic remains. Part 1. The Yorkshire Coast. 3d. ed. Edited by R. Etheridge. London, J. Murray. 354 pp.

QUENSTEDT, F. A.

1858 Der Jura. Tübingen, H. Laupp'sche buchhandlung. 842 pp.

SEELEY, H. G.

1880 On the skull of an *Ichthyosaurus* from the Lias of Whitby, apparently indicating a new species (*I. zetlandicus*, Seeley), preserved in the Woodwardian Museum of the University of Cambridge. Q. J. Geol. Soc. Lond., vol. 36, pp. 635-647.

WILSON, V.

1948 British regional geology; East Yorkshire and Lincolnshire. London, His Majesty's Stationery Office. 94 pp.

WOODWARD, A. S. AND C. D. SHERBORN

1890 A catalogue of British fossil Vertebrata. London, Dulau. 396 pp.

WOODWARD, H. B. AND W. A. E. USSHER

1911 The geology of the country near Sidmouth and Lyme Regis. 2d. ed. Mem. Geol. Surv. Engl. Wales. Sheet (B), nos. 326, 340, pp. 1-102.

ISBN 0-88854-151-1